THIS BOOK BELONGS TO:

_____

_____

© 2018 ATLANTIC JOURNALS - ALL RIGHTS RESERVED.

PUBLISHED BY ATLANTIC JOURNALS, 11923 NE SUMNER ST, STE 769907
PORTLAND, OREGON, 97220, USA

NO PART OF THIS PUBLICATION MAY BE REPRODUCED, STORED IN A RETRIEVAL
SYSTEM OR TRANSMITTED IN ANY FORM OR BY ANY MEANS, ELECTRONIC,
MECHANICAL, PHOTOCOPYING, RECORDING OR OTHERWISE, WITHOUT PRIOR
WRITTEN PERMISSION FROM THE AUTHOR/PUBLISHER.
FOR PERMISSIONS CONTACT: INFO@ATLANTICJOURNALS.COM

See Our Full Range At

ATLANTICJOURNALS.COM

© 2018 ATLANTIC JOURNALS - ALL RIGHTS RESERVED.

PUBLISHED BY ATLANTIC JOURNALS, 11923 NE SUMNER ST, STE 769907
PORTLAND, OREGON, 97220, USA

NO PART OF THIS PUBLICATION MAY BE REPRODUCED, STORED IN A RETRIEVAL SYSTEM OR TRANSMITTED IN ANY FORM OR BY ANY MEANS, ELECTRONIC, MECHANICAL, PHOTOCOPYING, RECORDING OR OTHERWISE, WITHOUT PRIOR WRITTEN PERMISSION FROM THE AUTHOR/PUBLISHER.
FOR PERMISSIONS CONTACT: INFO@ATLANTICJOURNALS.COM

See Our Full Range At

ATLANTICJOURNALS.COM

Made in the USA
Middletown, DE
02 September 2023

37820660R00113